YOUR KNOWLEDGE HAS VALUE

- We will publish your bachelor's and
  master's thesis, essays and papers

- Your own eBook and book -
  sold worldwide in all relevant shops

- Earn money with each sale

Upload your text at www.GRIN.com
and publish for free

Bibliographic information published by the German National Library:

The German National Library lists this publication in the National Bibliography; detailed bibliographic data are available on the Internet at http://dnb.dnb.de .

Imprint:

Copyright © 2015 GRIN Verlag, Open Publishing GmbH
Print and binding: Books on Demand GmbH, Norderstedt Germany
ISBN: 978-3-668-14990-8

This book at GRIN:

http://www.grin.com/en/e-book/315132/comparative-analysis-of-amo-breed-of-broiler-birds-fed-with-commercial

Nickson Ochika et al.

Comparative Analysis Of Amo Breed Of Broiler Birds Fed With Commercial And Self Formulated Feed

GRIN Publishing

COMPARATIVE ANALYSIS OF AMO BREED OF BROILER BIRDS FED WITH COMMERCIAL AND SELF FORMULATED FEED IN AGRICULTURAL SCIENCE EDUCATION DEPARTMENT OF FEDERAL COLLEGE OF EDUCATION, KATSINA STATE

By

NASIR MUSA

ABDULMUTALLAB ADAMU

NICKSON OJONUGWA OCHIKA

ADAMU SALE

CHINENYE NAOMI IWEOBI

MANSUR HASSAN

A project submitted to the school of vocational and Technical Education. Federal College of Education Katsina P.M.B 2041 Katsina, Katsina state.

In partial fulfillment for the requirement of the award of Nigeria certificate in education. (N.C.E)

November 2015.

Acknowledgement

The successful achievement of an individual cannot be attributed to himself alone. There are always people who either directly contributes in making ones dream come true. In this view, we want to show our gratitude to everyone who is one way or the other helped us to realize our dreams.

Firstly, our profound gratitude goes to almighty God who through his divine protection and guidance has made it possible for us to complete our study.

Secondly, we wish to express our sincere and profound gratitude to our supervisor Malam Mannir Sani for the assistance and encouragement and also for the time he took in reading through and correcting the manuscript, giving useful suggestions where applicable during the course of this research.

Also, we extend our profound gratitude to the entire lecturers of Agricultural department consisting of the Dean Mr Hassan Wutrama and the former Dean Mr Dauda Gambo, the head of Department Mallam Sa'idu Lawal, Dr Oscar T. Ikwakam, Engr Abubakar Yusuf, Mr. Agwam Yusuf, Mallam Manir Sani, Mallam Rabiu Abdullahi, Mallam Rabiu Abdullahi K/Bai, Mallam Auwal Mohammed, Mallam Aminu Ibrahim, and Mallam Yahaya Haruna and Lastly Mrs Adesina Adekambi.

We lastly wish to appreciate our parents, our siblings and also our friend for their support and moral encouragement during the course of my study and research work

Abstract

This study is aimed at comparing the effect of commercial and formulated feed on Amo breed of broilers. A comparative analysis of commercial and formulated feed on Amo breed of broiler was carried out.

Dedication

This practical research work is dedicated to God almighty for his protection and also to our parents for their financial support, prayers and moral encouragement.

Table of Contents

Chapter One

1.1   Introduction

The term "poultry" used in agriculture generally refers to all domesticated birds kept for eggs or meat production. These include chickens, (domesticated fowl) turkey, ducks, geese, pigeons, guinea fowl and other birds.

The most common of all these in Nigeria is the domestic fowl. Sometime the term "poultry" is considered synonymous with chickens (Igbokwe 2010).

The term poultry is a broad classification covering the species of all domesticated birds that are raised primarily for the production of meat and eggs for human consumption. The term applies to chickens, turkeys, ducks, swans, geese, guinea fowl and other birds.

However, among all these food producing birds, chickens are probably more than all the other birds and found throughout the world, (Feetham 1990).

Poultry production is one of the most important agricultural business practiced in Nigeria. But one of the challenges posed on the production of poultry is that the feed are mostly derived from crops such as maize, sorghum, and other grain crops. As a consequence, feed ingredients need to be acquired from distant supply area and may not be hundred percent (100%) sufficient in nutrients content and also very expensive (Nyoupayou 1990).

Researches have been undertaken since the 1930s on ways to improve feeds efficiency in broiler birds' production. Comparative trials have been conducted on the use of commercially formulated feeds and the locally produced ones to find out the most efficient form of feed that yields high growth rate and reduce production costs. (Mc Donald, 1987).

The meat and eggs obtained from chickens are good source of protein and fatty substances. There are both religious injunctions social taboos associated with pigs, consequently, chickens provide an acceptable form of animal protein to most people throughout the world (Smith, 1990).

Broiler birds are fast growing and specifically raised for meat, hence they are referred to as table birds. If properly managed, the birds have the capability to put on weight at such a rate that they are ready for market between 8-9 weeks when the live weight is 1.50 kg – 2.00 kg. (Akinsammi, 1994).

1.2   Statement of the problem

High cost of commercially prepared or factory feeds emanating from the cost and processing of the ingredients has resulted in declining productivity and profitability for intensive broiler production system. This scenario has resulted in supply bottle necks forcing an upsurge in the price of broiler birds in Nigeria.

Given the central importance that feeds play in intensive broiler system, it becomes imperative to identify the best type of feed options for the farmers.

1.3   Purpose of the Study

The research is purposely designed to:

1. Make a comparative analysis between Amo breed of broiler birds fed with commercial feeds and Amo breed fed with self-formulated feeds in F.C.E Katsina

1.4   Objective of the study

The main objective of this research project is to:

1. To determine the differences in terms of weight gain among the breed of birds as per the two types of feeds used on them.
2. To identify the effects of each feed to the live weight gain of the breed.

1.5   Significance of the study

This project is designed to make a comparative analysis between commercial feed and self-formulated feeds on single breed of broiler birds A1 (Amo) and A2 (Amo), where breed A1 (Amo) fed with commercial feed and A2 (Amo) fed with self-formulated feeds. The research will be of benefit to anybody who wants to set up a broiler production business. The findings of the study will also serve as reference material to students carrying out researches in similar areas.

1.6   Research Question

In an attempt to compare the effectiveness of the two different feeds on broiler birds, the following questions are investigated:

i.   What are the effects of commercial and self-formulated feeds on the weight of broiler birds?

ii.   Which of the feeds impacted positively on the birds?

1.7   Limitation of the study

This study is limited to an investigation of the effect of the two types of feeds: commercial feeds and self-formulated feeds on the performance of Amo breed broiler birds in the department of Agricultural Science Education, F.C.E Katsina.

Chapter Two

Literature Review

2.1   Introduction

The term "poultry" applies to chickens, turkey, ducks, geese, guinea fowl, and other game birds e.t.c.

Lexicon universal encyclopedia posits poultry as a branch of classification covering the species of domesticated birds that are raised to provide food to man. Banerjes, G.C (1992).

Feed ingredients needs to be acquired from distance supply areas which may not give hundred percent (100%) sufficient such as nutrient to the birds and of course they are expensive (Nyoupayou 1990).

Among all food producing birds, chickens, are probably, more than another single kind of birds and they live throughout the world. (Feetham 1990).

Trials have been done using commercial and self-formulated feeds to find out the most efficient form of feeds that yield high growth rate and reduce production cost M.C Donald (1987).

The meat and eggs obtain from chickens are good source of protein and fatty substance and all the products produce from chickens provides an acceptable form of animal protein to most people through the world (Smith 1990).

If properly managed, the birds have the capability to put on weight as such a rate that are ready for market between 8-9 weeks when the live weight is 1.50 kg – 2.00 kg (Akinsammi 1994).

Okere (1978), stated that, among all the farms animals, poultry can be a yield quicker return in a favorable condition and it is now believed that there is only a very few place on global where climatic condition makes the poultry flock impossible to be keeping.

He continued saying that poultry production in Nigeria can be established with minimum cost and apart from that they can be feed for themselves of free range without much care given to than or comprehensive management.

Okere (1993) also stated that poor feeders and drinkers result in wastage and contamination of feed. Contaminated feeds lead to spread diseases. The common example of this is coccidiosis.

Kesley (1980), said birds should be fed better from day old to about five weeks old. Finisher feeds ration should fed to the birds. The feed should be a high energy and protein ration formulation and hence promote rapid growth.

Rose el al (1986) covasa and forbes (1994), said about 70% of total farming cost in poultry production come from feed expenses. Many studies have been conducted to minimize this expenses. One way to minimize feed cost is to use raw material, from than at lower price just after harvest season. This potential should be used by poultry nutritionist to improve new feeding system or models. The ability of birds making their own diet has been used for this purpose in many studies.

The ability of domestic fowl to regulate their nutrient intake when allowed to choose between feeds containing high and low nutrients contents is well established. (Kauman et al 1978; Boorman 1999).

In the studies of two different feed having different nutrients content where specifically formulated or offered in plain form to birds.

However, there has been still a need to apply diet selection into the commercial environment with different application method (forbes 1995).

Broiler are birds main for meat production and that is the reason why carbohydrate and protein are been fed than so as to enable them put on flesh for maximum weight attainment (3 or 3.0 kg) not minimum weight (Henry 1986).

Any kinds of domestic animals not only birds, kept for rearing intensively (under intensive system) they need all necessary management to be given by the farmer to improve their proper growth and development and to be well productive. E.g poultry keeping. The broiler kept under the deep lither system or under battery cage system.

Under battery cage system, the birds are confined to individual cages these cages are built in house and adequate light and ventilation are provided to the birds. (Michael 1990).

Useful by product are obtain from poultry the feathers are used in stuffing pillows and mattress. The droppings contain more nitrogen, phosphorus and potassium in appreciable quantity than in those of other farm animals (okere 1993).

2.2   Live weight gain of birds as per the two type of feed used

Live weight gain is refer to as the weight of a living animal used for example as a pricing basis in selling livestock for meat. Live weight gain is the weight of an animal while living. It is the weight of an animal before it has been slaughtered and prepared as a carcass.

(Akinsanmi, 1991) opined that broilers are fast growing chickens which are specifically bred for meat production, hence and referred to as table birds. If properly managed, the breed has the capacity to put on weight at such a rate that they are ready for sale between 8-10 weeks of age when the live weight is about 1500g – 2000g.

2.3 Breed of broiler

Growing broilers-young chickens with pliable skin and tender meat involves making several choices. Your first decision is whether to involve making several choices. Your first decision is whether to raise hybrid or heritage poultry breeds. The fundamental differences are the amount of time they need to grow and the flavor of the meat. The quicker your birds reach the target weight of about 6 pounds the cheaper they are to raise overall and the more delicate the meat. The longer they take, the more they will cost you (as is the case with heritage breeds), but the meat will be healthier and more flavorful (Stein, 1996).

2.3.1 White hybrids:

The most efficient hybrid broiler chickens are an industrial creation developed by combining white cronish and white Plymouth Rock genetics. The resulting hybrids-the type most commonly sold at the same rate and efficiently convert feed into meat, reaching target weight in just six to seven weeks. Their edible portion (excluding excess fat, intestines, feathers, heads, feet and blood) is approximately 75 percent of live weight (Ravindran, 2013).

2.3.2 Colored hybrids:

The broilers were developed for France's famous label Rouge organic free-range chickens and adopted by some producers in the United States. Trade name includes Black Broiler, Color Yield, Colored Range Freedom Ranger, Kosher King, Redbro, Red Broiler, Red Meat Maker, Rosambro, and Silver Cross. Most strains have red plumage, but they also come in black, gray or barred, anything but white. Their colored feathers make them less visible to pluck cleanly so that the bare skin appears neat (Zhao et al, 2008).

Colored hybrid broiler chickens are usually raised on pasture and more slowly than white hybrids they take at least 11 weeks to reach target weight and the chicks do not necessarily grow at a uniform rate. They eat about 3 pound of feed per pound of weight gained because of their longer growth period and the calories they burn while foraging. Some people find the meat of colored hybrid to be more flavorful than that of faster growing white hybrids. The edible meat is approximately 70 percent of live weight (Zhao et al, 2008).

2.3.3 Heritage poultry breeds:

If you keep heritage chickens for eggs, you have the option of hatching eggs from your own flock, keeping the pullets as future layer and raising the surplus cockerels for meat.

All of these breeds are good forages and have a moderate to slow growth rate, reaching target weight in about 16 weeks. Compared with cornish hybrids, they have thinner breast and more dark meat. The higher foraging activity of heritage chickens results in meat that is lower in fat and farmer in texture. The meat has a richer chicken flavor because the birds are older when slaughtered *(Zhao et al, 2008).*

2.4   Feeding and feed conversion

The amount of feed that broiler chickens eat will depend upon the type of feed provided. Birds fed commercial-type diets will eat approximately twice as much feed as the amount of body weight gain.

As an example, birds processed for meat at 42 days of age may weigh 2.0kg(2000g) and have consumed 4.0kg(4000g) of feed. Chickens provided with a diet of lower   nutrient density will require more feed to reach a similar weight   (Shahidi et al, 1984).

Chickens are omnivores and modern broilers are given access to a special diet of high protein feed, usually delivered via an automated feeding system. This is combined with artificial lighting conditions to stimulate eating and growth and thus the desired body weight (Oluyemi and Robert 1979).

Diet is an important component of the environment under all climatic conditions. The potential of the bird cannot be attained if the environment, and notably the nutrition, is substandard (Oluyemi and Robert, 1979).

Generally, three types of feed are offered to broilers from day-old to marketing.

0.2 weeks   -   Broiler pre-starter mash

3-4 weeks   -   Broiler starter mash

5-6 weeks   -   Broiler finisher mash

2.5   Housing of Broiler

Broiler can be housed on deep-litter, splattered or wire floor or cages. However, cage, slat and wire floor rearing of broilers are not as popular as litter floor rearing, due to problems like breast blisters, leg   weakness and higher initial investment (Taylor, 1995).

Broilers can also be reared on cages. Broiler cages are similar to that of grower cages. To prevent the breast blister, the bottom of the cage may be coated with some plastic materials. The floor space requirement in cages is 15% of the floor space needed in deep-litter *(Taylor, 1995)*.

Broiler need good housing because of the following reasons:

- This is to protect birds against adverse weather condition

- It is also to protect birds against disease's attack
- It assists to keep birds in age groups for proper management
- It protects birds from thieves
- It protects birds from attack by wild animals, dangerous reptiles like snakes and from hawks
- Housing enhances maximum productivity in birds
- Good housing helps to increase the efficiency of feeding and feed utilization
- It facilitates management and veterinary care.

2.6 Handling and Transport of Broilers

The effects of handling and transportation of broilers are presented and discussed with particular reference to measure of the welfare of the birds. Point in the handling and transport process where improvements are required.

The procedures used to handle, transport spent hens, and broilers result in welfare problems for the birds, which are often very severe. There is evidence for substantial emergency responses, such as adrenal cortex activity. There can also be birds dead on arrival at the slaughter house, bruising, and high incidences of bone breakage. Housing broiler in battery cages for long period's results in bone fragility (Bayliss, 1990).

The greatest welfare problem is the normal rough handling which leads to poorer welfare than does a short vehicle journey. Better handling procedures are essential. Poor conditions on the journey also result in poor welfare.

Much research is needed using a wide range of indicators to assess welfare and to design housing systems, handling methods, and transport procedures which do not result in poor welfare (Bayliss, 1990).

2.7   Definition Of The Terms

- ❖ Broiler: chickens raised especially for meat purpose and marketed at about 2.0 kg live weigh.
- ❖ Feeders: are feeding materials e.g. rubber plats or plastic.
- ❖ Feeds: means the food given to poultry. {feeding containers}
- ❖ Drinkers: the material use for giving water to chickens, they also called waterier, {drinking containers}
- ❖ Friable materials: these are easily broken material used in raising poultry on the deep litter system example, saw dust.
- ❖ Feed ingredient: means each of the constituent materials making up a commercial feed or self-formulated feed.
- ❖ Ration: this is the quantity or the amount of feed given to bird a particular day/per one day.
- ❖ Deep litter system: this involves the floor of house covered straws dried grasses or wood and zinc roofing and must be concrete which absorbed the water associated with faces. And the birds are provided their needs in terms of water and feeds so also light and good ventilation.
- ❖ Mortality rate: this is the rate of death resulting from complete loss of life. The mortality rate of birds determine the amount number of birds dead.
- ❖ Broiler starter mash: is the feeds given to poultry birds {broiler} from four days to five weeks (4 days – 5 weeks) of age to point of disposal.
- ❖ Broiler finisher: is the feed poultry birds (broilers) from day old to five weeks of age.

Chapter Three

3.1   Research Methodology

This chapter covers all aspect of research methodology, which is the description of research work area ,method of data collection, sampling techniques and statistical tool used during the practical to ensure the more effective and accurate result from this research work.

3.2   Description of the research work area

In Nigeria, Katsina state was carved out from Kaduna state in the year 1987. The state consists of thirty four local government areas, the state consist of about 3,878,344 populations(1991 census) with total area of land of 24,192 km^2 (9,3415 sq ml), Katsina state is located in the north west zone of Nigeria, and the capital city is Katsina. The geographical location of the state is between latitude 12^0, 15^{0}N, 07^0, 30^{0}E and longitude 12^0, 250^0 N, 07,500^{0}E.

Among these local government, Batagarawa is included which was established in the year 1991 with total area of land of 433km^2 (167 sq ml) and the total number of population occupied the area of 184,575 (2006 census). It headquarters is located in the town of Batagarawa. The geographical area of Batagarawa local government is between latitude 12^0, 54^0,07^0,37^{0}E longitude 12, 900^0 N- 07, 617^{0}E. It has ten (10) total number of wards one out of these wards was where the research work took place, this ward named Batagarawa ward 'A' at Federal College Of Education.

The college was established in the year 1976 by the federal government of Nigeria. The geographical location is between Latitude 11^{0}N, 07^{0}N, and 13^0, 22^{0}N. Longitude 6, 52 W and 9, 42 E.

Agricultural Science education department is one of the pioneer departments of federal college of education, Katsina. The department started with establishment of the institution in 1978 at the temporary site in Dutsin-ma. Since the inception of the college, agricultural education department have been graduating students who have been carrying research work in different aspect of agricultural education.

In the past, student engage in literature kind of research work. However, in order to harness theory and practice, empirical research work is now being undertaken in the department group basis. The essence of this is to allow the students to practice what they have learnt and to come out with findings that will help in promoting agricultural science department.

3.3   Sampling size and Sampling Techniques.

This research work covers only the department of agricultural science education. A total population of (60) birds were use in this study and separated into two flocks (A^1 and A^2). Each flock contained (30) chicks. A^1 was Amo breed fed commercial feed and A^2 was Amo breed fed self-formulated feed. Random sampling technique was used in which 10 birds was randomly selected from each flock for the study.

3.4   Analytical tool and method of data analysis

Frequency table and percentages were analytical tools used to analyze the data collected.

3.5   Method of Data Collection

Observation and experimental methods were used in gathering and collecting this data's. The daily feed intake was collected on daily basis while the weekly weight gain was determined on weekly basis. The data collection lasted for 8 weeks. The data used for the study were made up of the daily feed consumption of the poultry birds and their weekly weight gain. Initial weight of the poultry birds was recorded at a day old before the commencement of the feeding trial. The weight gained by the birds was determined by a random weighing of thirty (10) birds in each flock. This gave a total of sixty (20) birds randomly selected for weighing weekly. Feed consumption of the birds was also recorded. Both the weight gained and the quantities of feed consumed by the birds were in gram. For determining the total weight gained of the birds, weekly weight gained was recorded.

Chapter Four

4.1   Results and Discussions

Below are the results obtain from the research and are discussed below.

4.1.1 Average Weekly Feed Consumption of the poultry birds

Table 1 shows the result of the average weekly feed consumption in gram per poultry as they were fed with two types of feed (commercial and self-formulated feed) and the percentage of each feed per week.

The feed intake is calculated by:

Feed intake = feed given – left over

The percentage is calculated by = $\underline{\text{feed intake per week}} \times 100$

Total feed intake in 8 weeks

Table 1: average weekly feed consumption in gram per poultry in each flock.

Weeks	Feed intake of A^1 in grams	Feed intake of A^2 in grams	Feed percentage of A^1	feed percentage of A^2
1	77.3	62	2.16%	3.09%
2	228	100.7	6.36%	5.02%
3	305.3	135	8.52%	6.68%
4	533.3	105	14.9%	5.24%
5	636.7	235.7	17.8%	11.76%
6	358.2	165.7	9.9%	8.27%
7	611.3	200	17.08%	9.98%
8	833.3	666.7	23.25%	33.28%
Total	3584g	2003g		

As the poultry birds increase in their age, their weekly feed consumption was increasing from the first week up to eight (8) week. The initial weight of the A^1 flock (Amo breed fed commercial feed) was (18.14) g at a week old , while that of the A^2 flock (Amo breed fed self-formulated feed) was (16.34)g. From week one to week eight, the Amo breed A^1 obtained a weight of (770.11) g, while the Amo breed A^2 obtained a weight of (585.6)g.

4.1.2 The weight gain of the poultry

The weight gains of the poultry were recorded on weekly basis. Table 2 shows the weekly weight gain of the poultry and the percentage of weight obtain per week.

$$\text{The weight gain is calculated by} = \frac{\text{sum of the weight of birds sample}}{\text{Number of birds sampled}}$$

The percentage of weight gain =

$$\frac{\text{sum of the weight of birds sweight gain per week} \times 100}{\text{Total weight in 8 weeks}}$$

Table 2: the average weekly weight gain of the poultry in gram.

Weeks	Weight gain of A^1 in gram	Weight gain of A^2 in gram	Weight percentage in A^1	Weight percentage in A^2
1	18.14	16.34	0.7%	0.9%
2	125.5	38.19	4.6%	2.3%
3	144.7	55.3	5.3%	3.3%
4	253	94.6	9.2%	5.7%
5	361.3	194.6	13.1%	11.7%
6	462.04	285.6	16.9%	17.2%

7	603.7	385.6	22.04%	23.2%
8	770.4	585.6	28.1%	35.34%
Total	2738.9g	1656.9g		
Average	273.89g	165.69g		

The table above (table 2) shows the weight of the poultry in flock A^1 and A^2. The weight of the broilers in the two different flocks continue to increase in weight and percent until the eight week where the highest weight gained by the birds in flock A^1 was 770.4g and that of the flock A^2 was 585.6g. The total weight gained at the end of the research was 2738.9g for the flock A^1, while that of the flock A^2 was 1656.9g. The average weight gain was obtained by adding up the weekly weight gain of the different birds and the result divided by the total number of poultry measured. Below show how the average weekly weight gain was obtained.

Below is the average weekly weight gain of flock $A^1 =$

$$\frac{2738.9}{10} = 273.89 \; kg$$

Below is the average weekly weight gain of flock $A^2 =$

$$\frac{1656.9}{10} = 165.69 \; kg$$

4.1.3 The nutrient constituent of the self-formulated feed

The below table 3 show the nutrient constituent of the self-formulated feed

S/N	Items	DES	Quantity	Price/unit	Amount
1	Maize	Feed component	3 bags	₦7100	₦21,700
2	Cone	Feed component	3 bags	₦3850	₦15,400
3	Bone meal		20 p	₦150	₦3,000
4	Pre-mix		2 sachet	₦1000	₦2000
6	Wheat Bran		3 bags	₦3000	₦9000
Total				₦15,100	₦51,100

Above are the list of nutrient needed to formulate a poultry feed. The commercial feed use is the Vital feed. The total price for formulating feed is ₦51,100. A total of 15 bags of commercial feed was used (vital feed) for the research which cost ₦42,720. While the transport used is ₦2000.

4.1.4 Discussions of Findings

The analysis of data collected in this research is discussed as follows:

Table 1 shows that the Amo breed A^1 (Amo breed fed commercial feed) consumed more feeds than the Amo breed A^2 (Amo breed fed self-formulated feed).

Table 2 shows that the weight of the Amo breed A^1 (Amo breed fed commercial feed) are more than the Amo breed A^2 (Amo breed fed self-formulated feed). This is because the A^1 flock broilers birds consume more food than A^2 flock broilers birds.

Table 3 shows the nutrient constituent of self-formulated feed. From the table, the price for formulating a poultry feed is ₦51,100 which tend to be higher than the price of the commercial feed used. The formulated feed was   used in flock A^2 while the commercial feed was used in flock A^2.

Chapter Five

5.1   Summary, Conclusion and Recommendations

The Study was carried out on the comparative analysis between Amo breed of broiler birds fed with commercial feeds and Amo breed fed with self-formulated feeds in F.C.E Katsina. The data for the research work were from primary source collected on daily feed intake and the weekly weight gain of the broiler birds.

The two different feeds were fed on the same proportion to the birds from the time they were bought at a day old until they were eight weeks old. The research work specific objectives are to determine the differences in terms of weight gain among the breed of birds as per the two types of feeds used on them.

5.2   CONCLUSIONS

Based on the result obtained from this research work conducted, it shows that the commercial feed has a higher growth rate than the self-formulated feed. The only noted significant differences between the two feed is in term of the price. The formulated feed cost ₦51,100 to formulate while the commercial feed cost ₦42,720.

5.3 RECOMMENDATIONS

The study shows that commercial feed has more productive characteristics than the self-formulated feed under the same feeding proportion. Based on the findings in this research, the following recommendations were made:

1. Broiler producers should adapt the use of commercial feed over the self-formulated feed in broiler production.
2. More research should be done on how to continue improving the self-formulated feed. And maintaining and perfecting the commercial feed.

3. From the result, it is important to note that in selecting feed, commercial feed is the best because of the weight attain by the birds in flock A^1 fed with it and thus it will yield high profit.

REFERENCES

Adebowale, E.A., A.M. Bamgbose and Nworgu F.C, (1998). Performance of
Broilers fed different Protein Sources. Proc. Silver Ann. Conf. of Nig.
Soc. For Ani. Prod., 21-25 march, 1998, Abeokuta, Nigeria, pp: 596-597.

Agbamu,J.U., (2005). Problems and Prospects of Agricultural Extension
Service in
developing ccountries. Agric. Exten. Soc. Nigeria, Ilorin, Nigeria.

Ahuja, Vinod 2007 Scope and Space for small scale poultry production in
developing countries

ALDERS et al (1977) Village chicken production in Bilene District,
Mozambique: Current practices and problems. Paper presented at 1NFPD
Workshop and   General Meeting, 9-13 December 1997, M'Bour, Senegal

Alders, R.G. & Spradbrow, P.B. (2000). Introduction. *In* R.G. Alders & P.B.
Spradbrow, eds. SADC Planning Workshop on Newcastle Disease
Control   in Village Chickens. Maputo, Mozambique, *6-9 March.* Canberra,
Australian Centre for International Agricultural Research.

Astroth, K.A., & Robbins, B.S. (1987). Recess is over. Journal of Extension,
25(3) (Online). Available: http://www .joe.org/joe/1987fall/a2.html. Accessed:
01/15/02.

Amir, H.N., Y. Mojtaba and D.B. Gary, (2001). How nutrition affects immune
responses in poultry. World Poult. Elsevier, 17: 6. Esonu, B.O, (2000).

Animal Nutrition and Feeding: a functional approach. Published by Rukzeal   and Ruksons Associates Memory Press, Owerri, Imo State, Nigeria.

Astroth, K.A., & Robbins, B.S. (1987). Recess is over. Journal of Extension, 25(3) (Online). Available: http://www .joe.org/joe/1987fall/a2.html. Accessed: 01/15/02.

B. Sonaiya (2007) Toward sustainable poultry production in Africa

Basnyat, B.B. (1990). Agricultural Extension System in Nepal, Development

Buyse,J. Simons, P.C.M, Boshouwers, F.M.G. Decuypere, E. (1996). Effects of intermittent lighting, light intensity and source of the performance and welfare of broilers. World's poultry science journal 52:121-130.

Jones, T.P., (2005). Quality control in feed manufacturing, Avitech Technica bulletin, http://www.thepoultrysite. com/articles/526/quality-control-in feed-manufacturing. Assessed on 17th July, 2007.

Okoli, I.C., A. A., Omede, I.P. Ogbuewu and M.C. Uchegbu, (2009). Physical Characteristics As Indicators Of Poultry Feed Quality: A Review. Proceedings of the 3rd Nigeria International Poultry. Summit. S.I., Ola., Fafiolu, A.O. and Fatufe, A.A. (Edn). 22-26 February 2009. Abeokuta, Ogun State, Nigeria, pp: 124-128.

McDonald, D. (1987). Animal Production, 3rd Edition; Longman New York, U.S.A.

Mutetwa, L (2001). Irvin"s National Foods and ZIMVET Newsletter Vol 8, Number 2.

Nkomo, N (2001). Irvin"s National Foods and ZIMVET Newsletter Vol 8, Number 2.

Nyoupayou, J.D.N. (1990). Country Report of Small Rural International on Rural Poultry Production; Thessalonik, Greece.